My Life Book

If you purchase this book without a cover you should be aware that this book may have been stolen property and reported as "unsold and destroyed" to the publisher. In such case neither the author nor the publisher has received any payment for this "stripped book."

Copyright © 2008 by Vickie Smith

All rights reserved. Except as permitted under the U.S. Copyright © 1976, no part of this publication may be reproduced, distributed, or transmitted in any from or by any means, or stored in a database or retrieval system, without the prior written permission of the publisher

Cover designed by Vickie Smith

Printed in the USA

Lulu item 2275800
ISBN 978-0-6152-0221-1

Published by:
Vickie C. E. Smith

PastorsSmith@sbcglobal.net
Indianapolis, IN

My

Life

Book

Name_____   Date_____

Presented to

_____

By

_____

On the Occasion of

_____

Date

Extra Special Occasions To Give The My Life Book

Baby's birth

Weddings

Family Reunions

Graduations

Christmas

Welcome to your life book; be sure to read each chapter from beginning to end before you start your entries. Some of the questions may not pertain to you as this book has been written for every ones uses. Make your life book your own, leave important information for your family and friends to remember you just the way you are. This book will allow generation after generation to know their ancestors.

My Life Book can be used in many ways,

1. Anyone can start a *My Life Book* for the new beginning of life when a couple comes together as one at a wedding.
2. Give a *My Life Book* for the new beginning of life when a couple comes together as one at a wedding.
3. Start the *My Life Book* at your family reunion where each person can fill in the names of their ancestors, then have fun recording stories of the past for their children to enjoy.
4. Give the *My Life Book* as a Graduation gift for the new graduate to record the steps they will take as they start their new journey in life.
5. Make it a Christmas to remember when it begins with the story of their life by giving a *My life Book*.

The *My Life Book* is just that, a book that allows you to tell your life in your own words. The *My Life Book* allows you to leave your thoughts, ideas and beliefs to those that will not know you if they are born after your death. *My Life Book* was created to give you a chance to say all the things you want to say, or could not say while you were alive. *My Life book* can stand up for you when you pass away and someone tries to say something about you that's not true, your *My Life Book* can stand up and speck for you when you are gone. *My Life Book* is a great way for a parent to leave the memories of their life for their children and love ones. *My Life Book* is a wonderful way for great and grand-parent's to leave the memories of their life for their great and grand-children. If you have anyone in your family or know of someone that may be showing signs of Alzheimer, please start a *My Life Book* for them. And please take the time to read their stories to them as often as possible.

Contents

My Family's Beginning..Page 15
    Country
    Addresses
    Neighbors

Who I Am...16
    Family Tree
    Non Family
    Adoptions

My Relationship With My Father...20
    His Love
    That Special Moment(s)
    The Good
    The Bad
    The Ugly

My Relationship With My Mother..30
    Her Love
    That Special Moment(s)
    The Good
    The Bad
    The Ugly

My Beliefs..40
    In Life
    In People
    In Family
    In God (or not)
    In My Country

About My Spouse(s)...50
    Day 1
    Proposal(s)
    Wedding(s)
    Divorce(s)

Health...61
    Family History
    Allergies
    Surgeries
    Diseases

About My Life Partner...63
    Day 1
    That Special Moment(s)
    Why We Chose Not To Get Married
    Why We Could Not Get Married

My Children..70
    Birth
    Non Birth
    Adoptions
    Abortions

My In-laws..74
    The Good
    The Bad
    And The Ugly

My Friends...86
    Best
    True
    False
    Worst

My Goals In Life...108
    First
    Accomplished
    Unreachable

Education..114
    Grade School
    Middle School
    High School
    Collage School
    Grad School

Work History..125
    First
    Promotions
    Retirement

Volunteer Work..131
    Community
    Church
    World

My Dreams..137
    For Self
    For Spouse(s)
    For Children
    For Family
    For Friend(s)
    For Others
    For the World

My Fears..151
    Fear of Life
    Fear of People
    Fear of Success
    Fear of Failure
    Fear of Things

My Hope(s)..161
    For Self
    For Spouse(s)
    For Children
    For Family
    For Friend(s)
    For Others
    For the World

Time I Wasted...175
    For The Better
    The Worst
    The Most
    That I Regret

People That Are Special To Me..183
    Mentors
    Teachers
    Spouse(s)
    Family
    Friend(s)
    In the World

People that Changed My Life..195
    For Good
    For Bad
    Other

An Event(s) That Changed My Life...201
    For Good
    For Bad
    Other

Places I Have Lived..207
    Love
    Hate
    Wish I Was Still Their

Things I Invented..213
    My Idea
    My Invention
    Ideas I Have Patented

Lies I Have Told..219
    Little White
    To Self
    To Family
    To Friends
    To Spouse
    To The Law
    About Money
    Because I Felt I had No Other Choice

The Lie(s) I Have Lived..235
    About Self
    For My Family
    For A Friend
    About Money
    For Image Sack
    Because I Felt I had No Other Choice

My Most Embarrassing Moment(s)..245
    In Life
    In School
    With Family
    At Work
    In Public
    That Was Out Of My Control
    That Was Within My Control

Party..259
    Never
    All The Time
    With Family
    W Friends

When I Was Jealous..267
    Of Spouse(s)
    Of Children
    Of Family
    Of Friend(s)
    Of Others

When I Thought Others Were Jealous Of Me..277
    My Spouse(s)
    My Children
    My Family
    My Friend(s)
    Others

Vacation(s)..287
    Alone
    With Family

My Likes & Dislikes With..291
    My Life
    My Spouse(s)
    My Children
    My Family
    My Friend(s)
    My Job
    With Others
    With The World
Politics..307
    Democrat
    Republican
    Voting
Hobbies...313
    Skills
    Collections
    Collectables
That Thing I Always Wanted To Say...320
    To Family
    To Friends
    To Spouse(s)
    To the Law
    To The World
That Thing I Always Wanted To do..330
    With Family
    With Friends
    With Spouse(s)
Taxes..338
    How I Feel
    How I Believe
    What I Think Should Be Done
Racism...344
    The Good
    The Bad
    And The Ugly
Computers..350
    Computers In My Life
    Computers In The World
How I Feel About Myself As A..354
    Man
    Woman
Military Service...360
    Service
    Rank /Medal(s)

How I Feel About Myself As A 364
    Sister
    Brother
    Cousin
    Aunt
    Uncle
    Grandmother
    Grandfather
    Great-grandmother
    Great-grandfather

Homosexuality 390
    What I Understand
    What I Believe

Sports 394
    My Achievements
    My View On Professional Sports

My View On Global Warming 398
    Believe In
    Disbelieve

My View On Crime 402
    What I Think
    What I Know
    What I believe

Book(s) 408
    Read
    Recommend

Vehicles 412
    First
    Wanted
    Love

Television 418
    Favorite
    Worst
    Love

Actor - Actress 424
    Favorite
    Worst
    Love

Outer Space 430
    Space Travel
    Aliens
    What I Believe

Music 436
    Love
    Hate

Movie(s)..440
    Favorite
    Worst
    Love
    Hate
Shopping..448
    Love
    Hate
    Addicted
Favorite Designer(s)...454
    My Favorite Dress - Suite
    My Favorite Shoe or Handbag
Favorite Weather..458
    Winter
    Spring
    Summer
    Fall
My Favorite Food...466
    Favorite
    Worst
    Love
    Hate
My Pet(s)...474
    Dog(s)
    Cat(s)
    Other
My View On Life Support...480
    For Myself
    For Others
    What I believe
My View On The Death Penalty..484

My View On Death...486
    For Myself
    For Others
    What I believe
In My Own Words..494

What I Think Will Happen After I Am Gone..506
    With Family
    With The World
My Last Will And Testament..512
Notary...516

Place photo here

*Name*_____

*Address*_____

*City*_____ *St.*____ *Zip.*____

I Leave My Life Book To:

_____

My Family's Beginning

Try to give as much information as you can about where your family came from or began.

Country

_____
_____
_____
_____
_____
_____
_____
_____

Addresses

Try to give as many addresses as you can to help with any research your love ones may want to do on the family.

_____
_____
_____
_____
_____
_____
_____
_____

Neighbors

Try to give as many neighbors names and addresses as you can to help with any research your love ones may want to do on the family history.

_____
_____
_____
_____
_____
_____
_____
_____
_____

Who I Am
Family Tree
Starting with the oldest to the youngest person I know in my family.

Adoptions ## *Non Family*

Starting with the oldest to the youngest person Adopted or like family with no bloodline I know in my family.

Use this page to explain any questions that you feel my come up after your life book is read concerning, Adopted or Non Family members.

My Relationship With My Father

His Love

Tell about the special love your father gave you.

Today's date_____

Tell about that special moment(s) you shared with your father.

That Special Moment(s)

Today's date_____

The Good

Tell about the good things you love about your father.

Today's date _____

The Bad

Tell about the bad things you remember about your father.

Today's date_____

The Ugly

Tell about the ugly things you want to share about your father.

Today's date_____

My Relationship With My Mother

Her Love

Tell about the special love your mother gave you.

Today's date_____

The Good

Tell about the good things you love about your mother.

Today's date _____

The Bad

Tell about the bad things you remember about your mother.

Today's date_____

And The Ugly

Tell about the ugly things you want to share about your mother.

Today's date _____

My Beliefs
Tell about your belief, and what they mean to you.

In Life

Today's date_____

In People

Tell about your beliefs in other people.

Today's date_____

In Family

Tell about your beliefs in your family.

Today's date _____

Tell about your beliefs in God or not.

In God (or not)
Church Name _____
Pastor _____ First Lady _____
Address _____ State _____ City _____

I served as a
Check all that apply

Bishop[ ]   Pastor[ ]   Assistant Pastor[ ]   Minster[ ]   Evangelist[ ]
Usher[ ]   Choir Member[ ]   Pastor Aid[ ]   Volunteer[ ]   Missionary[ ]

Today's date _____

In My Country

Tell about your beliefs in your country.

Today's date _____

About My Spouse(s)

Tell about the first day you met your spouse.

Day 1

Today's date_____

List all proposals you have had in your life time, starting with the first to the last.

Proposal(s)

Proposal 1.   Date_____   Person_____
                        Accept   Yes[ ]   No[ ]

_____

Proposal 2.   Date_____   Person_____
                        Accept   Yes[ ]   No[ ]
                   Today's Date_____

_____

Proposal 3.   Date_____   Person_____
                    Accept   Yes[ ]   No[ ]
              Today's Date_____

_____

Proposal 4.   Date_____   Person_____
                    Accept   Yes[ ]   No[ ]
              Today's Date_____

_____

Proposal 5.　　Date_____　　Person_____
　　　　　　　　　　Accept　　Yes[ ]　　No[ ]
　　　　　　　Today's Date_____

Proposal 6.　　Date_____　　Person_____
　　　　　　　　　　Accept　　Yes[ ]　　No[ ]
　　　　　　　Today's Date_____

List all Weddings you have had in your life time, starting with the first to the last.

Wedding(s)
Wedding 1.   Date_____   Person_____
Yes[ ]   No[ ]
Today's Date_____

_____
_____
_____
_____
_____
_____
_____
_____
_____
_____
_____
_____
_____
_____
_____
_____
_____
_____
_____

Wedding 2.   Date_____   Person_____
Yes[ ]   No[ ]
Today's Date_____

_____
_____
_____
_____
_____
_____
_____
_____
_____
_____
_____
_____
_____
_____
_____
_____
_____

Wedding 3. Date_____ Person_____
Yes[ ]   No[ ]
Today's Date_____

Wedding 4. Date_____ Person_____
Yes[ ]   No[ ]
Today's Date_____

Wedding 5.   Date_____   Person_____
                          Yes[ ]   No[ ]
                Today's Date_____

Wedding 6.   Date_____   Person_____
                          Yes[ ]   No[ ]
                Today's Date_____

Divorce(s)

List all Divorces you have had in your life time, starting with the first to the last.

Divorce 1. Date_____ Person_____
Yes[ ] No[ ]
Today's Date_____

_____
_____
_____
_____
_____
_____
_____
_____
_____
_____
_____
_____
_____
_____
_____
_____
_____
_____
_____
_____

Divorce 2. Date_____ Person_____
Yes[ ] No[ ]
Today's Date_____

_____
_____
_____
_____
_____
_____
_____
_____
_____
_____
_____
_____
_____
_____
_____
_____
_____
_____
_____
_____

Divorce 3.   Date_____   Person_____
                          Yes[ ]   No[ ]
                  Today's Date_____

_____
_____
_____
_____
_____
_____
_____
_____
_____
_____
_____
_____
_____
_____
_____
_____
_____
_____
_____
_____

Divorce 4.   Date_____   Person_____
                          Yes[ ]   No[ ]
                  Today's Date_____

_____
_____
_____
_____
_____
_____
_____
_____
_____
_____
_____
_____
_____
_____
_____
_____
_____
_____
_____
_____

Divorce 5. Date_____ Person_____
　　　　　　　　　　　Yes[ ]　　No[ ]
　　　　　　　　Today's Date_____

Divorce 6. Date_____ Person_____
　　　　　　　　　　　Yes[ ]　　No[ ]
　　　　　　　　Today's Date_____

Health

List all important health issues that need be addressed for others that may have concern about their own health.

Family History
Father's Side   Today's Date _____

_____
_____
_____
_____
_____
_____
_____
_____
_____
_____

Mother's side

_____
_____
_____
_____
_____
_____
_____
_____
_____
_____

Allergies
Father's Side   Today's Date _____

_____
_____
_____
_____
_____
_____
_____
_____
_____
_____

Mother's side

_____
_____
_____
_____
_____
_____
_____
_____
_____
_____

Surgeries
Father's Side

Today's Date _____

Mother's side

Diseases
Father's Side

Today's Date _____

Mother's side

About My Life Partner

Day 1

Today's Date_____

That Special Moment(s)

Today's Date _____

Why We Chose Not To Get Married
Today's Date_____

Why We Could Not Get Married

Today's Date_____

My Children

List all your children from first to last in order of birth.

Birth Children

Today's Date _____

Date 1. _____ Name _____

Date 2. _____ Name _____

Date 3. _____ Name _____

Date 4. _____ Name _____

Date 5. _____ Name _____

Date 6. _____ Name _____

Date 7. _____ Name _____

Date 8. _____ Name _____

Date 9. _____ Name _____

Date 10. _____ Name _____

Date 11. _____ Name _____

Date 12. _____ Name _____

Date 13. _____ Name _____

Date 14. _____ Name _____

Date 15. _____ Name _____

Date 16. _____ Name _____

Adoptions

List all children you have adopted from first to last

Today's Date_____

Birth Date   Full Name

Date 1._____ Name _____

Date 2._____ Name _____

Date 3._____ Name _____

Date 4._____ Name _____

Date 5._____ Name _____

Date 6._____ Name _____

Date 7._____ Name _____

Date 8._____ Name _____

Date 9._____ Name _____

Date 10._____ Name _____

Date 11._____ Name _____

Date 12._____ Name _____

Date 13._____ Name _____

Date 14._____ Name _____

Date 15._____ Name _____

Date 16._____ Name _____

List all children you did not give birth to that you call your child
(Foster children, Children you have taken in and raised as your own not through adoption.)

Non Birth

Today's Date_____

Birth Date   Full Name

Date 1._____Name _____City_____St____

Date 2._____Name _____City_____St____

Date 3._____Name _____City_____St____

Date 4._____Name _____City_____St____

Date 5._____Name _____City_____St____

Date 6._____Name _____City_____St____

Date 7._____Name _____City_____St____

Date 8._____Name _____City_____St____

Date 9._____Name _____City_____St____

Date 10._____Name _____City_____St____

Date 11._____ Name _____City_____St____

Date 12._____ Name _____City_____St____

Date 13._____ Name _____City_____St____

Date 14._____ Name _____City_____St____

List all children you did not give birth to but aborted from first to last.

Abortions

Today's Date_____

Abortion Date	Father/Mother		
Date 1._____	Name _____	City_____	St____
Date 2._____	Name _____	City_____	St____
Date 3._____	Name _____	City_____	St____
Date 4._____	Name _____	City_____	St____
Date 5._____	Name _____	City_____	St____
Date 6._____	Name _____	City_____	St____
Date 7._____	Name _____	City_____	St____
Date 8._____	Name _____	City_____	St____
Date 9._____	Name _____	City_____	St____
Date 10._____	Name _____	City_____	St____

My In-laws

Tell all the good things about your father-in-law you want your love ones to know.

The Good
About my Father-in-law

Today's Date_____

Tell all the good things about your Mother-in-law you want your love ones to know.

The Good
About my Mother-in-law
Today's Date_____

Tell all the bad things about your father-in-law you want your love ones to know.

The Bad
About my Father-in-law

Today's Date_____

Tell all the bad things about your mother-in-law you want your love ones to know.

The Bad
About my Mother-in-law

Today's Date_____

Tell all the ugly things about your father-in-law you want your love ones to know.

And The Ugly
About my Father-in-law

Today's Date_____

Tell all the ugly things about your mother-in-law you want your love ones to know.

And The Ugly
About my Mother-in-law

Today's Date_____

My Friends

Tell all the good things about your best friend you want your love ones to know.

Best Friend

Today's Date_____

Best Friend

Tell all the bad things about your best friend you want your love ones to know.

Today's Date_____

Best Friend

Tell all the ugly things about your best friend you want your love ones to know.

Today's Date_____

True Friend

Tell all the good things about your true friend(s) you want your love ones to know.

Today's Date_____

True Friend

Tell all the bad things about your true friend(s) you want your love ones to know.

Today's Date _____

True Friend

Tell all the ugly things about your true friend(s) you want your love ones to know.

Today's Date_____

Tell all the good things about your false friend(s) you want your love ones to know.

False Friends

Today's Date_____

Tell all the bad things about your false friend(s) you want your love ones to know.

False Friends

Today's Date _____

Tell all the ugly things about your false friend(s) you want your love ones to know.

False Friends

Today's Date_____

Worst Friends

Tell all the good things about your worst friend(s) you want your love ones to know.

Today's Date_____

Tell all the bad things about your worst friend(s) you want your love ones to know.

Worst Friends

Today's Date_____

My Goal(s) In Life

Give in detail the goal(s) you set for yourself, from the first to the last.

First

Today's Date _____

Give in detail the goal(s) you have accomplished from the first to the last.

Accomplished

Today's Date_____

Give in detail the goal(s) you felt were unreachable from the first to the last.

Unreachable

Today's Date_____

Education

Today's Date_____

Grade School

Name_____

Address_____

City_____ St._____ Zip._____

Passed YES [ ]   NO [ ]

Middle School

Name_____

Address_____

City_____ St._____ Zip._____

Passed YES [ ]   NO [ ]

High School

Name_____

Address_____

City_____ St._____ Zip._____

Graduate YES [ ]   NO [ ]

Collage

Name_____

Address_____

City_____ St._____ Zip._____

Graduate YES [ ]   NO [ ]

Grad School

Name_____

Address_____

City_____ St._____ Zip._____

Graduate YES [ ]   NO [ ]

Grade School

Tell about your experience in grade school.

Today's Date_____

Middle School

Tell about your experience in middle school.

Today's Date_____

High School

Tell about your experience in high school.

Today's Date _____

Collage

Tell about your experience in collage.

Today's Date_____

Tell about your experience in grad school.

Grad School

Today's Date _____

Work History

Tell about the job(s) you have held in your life time.

First

Today's Date_____

Promotion(s)

Tell about the promotion(s) you have received in your life time.

Today's Date_____

Retirement

Give your thoughts on retirement.

Today's Date_____

Volunteer Work
Tell about your experience as a volunteer in your community.

Community

Today's Date_____

Tell about your experience as a volunteer in the church.

Church

Today's Date_____

Tell about your experience as a volunteer in the world.

World

Today's Date_____

My Dreams

Tell about the dream(s) you have or had for yourself.

For Self

Today's Date _____

For Spouse(s)

Tell about the dream(s) you have or had for your spouse(s)

Today's Date_____

For Children

Tell about the dream(s) you have or had for your children.

Today's Date_____

Tell about the dream(s) you have or had for your family, this could be for other family members that are not your children.

For Family

Today's Date_____

Tell about the dream(s) you have or had for your friends, or what you believe would have been a dream come true for a friend(s).
For Friend(s)

Today's Date_____

Tell about the dream(s) you have or had for others, or what you believe would have been a dream come true for others.

For Others

Today's Date_____

Tell about the dream(s) you have or had for the world, or what you believe would have been a dream come true for the world.

For the World

Today's Date_____

My Fears

Tell about the fear(s) you have or had about life.

Fear of Life

Today's Date _____

Fear of People

Tell about the fear(s) you have or had of other people.

Today's Date_____

Fear of Success

Tell about the fear(s) you have or had about success.

Today's Date_____

Fear of Failure

Tell about the fear(s) you have or had about failure.

Today's Date_____

Fear of Things

Tell about the fear(s) you have or had of things.

Today's Date_____

My Hope(s)

Tell about the hope(s) you had for yourself.

For Self

Today's Date_____

For Spouse(s)

Tell about the hope(s) you had for spouse(s).

Today's Date_____

For Children

Tell about the hope(s) you had for your children.

Today's Date _____

For Family

Tell about the hope(s) you had for your family.

Today's Date_____

For Friend(s)

Tell about the hope(s) you had for your friends.

Today's Date_____

For Others

Tell about the hope(s) you had for others.

Today's Date_____

For the World

Tell about the hope(s) you had for the world.

Today's Date_____

Time I Wasted For The Better

Tell about the time you wasted that turned out to be for the better.

Today's Date_____

The Worst

Tell about the time you wasted that turned out for the worst.

Today's Date_____

The Most

Tell about the most time you wasted.

Today's Date _____

That I Regret

Tell about the time you regret wasting.

Today's Date_____

People That Are Special To Me

Tell about the person you call your mentor(s).

Mentor(s)

Today's Date_____

Teachers

Tell about the teacher(s) you feel were special to you.

Today's Date_____

Spouse(s)

Tell about the spouse(s) you feel were special to you.

Today's Date _____

Family

Tell about the family member(s) you feel were special to you.

Today's Date _____

Friend(s)

Tell about the friend(s) you feel were special to you.

Today's Date_____

In the World

Tell about something you feel that was special to you in the world.

Today's Date_____

People that Changed My Life For Good

Tell about the people that changed your life for the good.

Today's Date_____

For Bad

Tell about the people that changed your life for the bad.

Today's Date _____

Other

Tell about the people that changed your life in other ways.

Today's Date_____

An Event(s) That Changed My Life For Good

Tell about an event(s) that changed your life for the good.

Today's Date_____

For Bad

Tell about an event(s) that changed your life for the bad.

Today's Date_____

Other

Tell about an event(s) that changed your life for other reasons.

Today's Date_____

Loved

Places I Have Lived
Tell about that place were you lived and you loved living there.

Today's Date_____

Hated

Tell about that place were you lived and you hated living there.

Today's Date _____

Tell about that place were you lived and you wish you still lived there.

Wish I Was Still Their

Today's Date_____

Things I Invented

Tell about the idea(s) you have or had.

My Idea(s)

Today's Date_____

My Invention(s)

Tell about the invention(s) you created.

Today's Date _____

Idea(s) I Have Patented

Tell about the invention(s) you have patented.

Today's Date_____

Lies I Have Told

Tell about the little white lies you have told.

Little White

Today's Date_____

To Self

Tell about the lies you have told to yourself.

Today's Date_____

To Spouse(s)

Tell about the lies you have told to your spouse(s).

Today's Date_____

To Family

Tell about the lies you have told to your family.

Today's Date_____

To Friends

Tell about the lies you have told friends.

Today's Date _____

To the Law

Tell about the lies you have told to the law to protect yourself or others.

Today's Date_____

About Money

Tell about the lies you have told concerning money.

Today's Date_____

T

Tell about the lie(s) you have told that you felt you had no choice.
Because I Felt I had No Other Choice
Today's Date_____

The Lie(s) I Have Lived

Tell about the lie(s) you have lived that no one knows about.

About Self

Today's Date_____

For My Family

Tell about the lie(s) you have lived for your family that no one knows about.

Today's Date _____

For A Friend

Tell about the lie(s) you have lived for a friend that no one knows about.

Today's Date_____

For Image Sack

Tell about the lie(s) you have lived for the sack of your image that no one knows about.

Today's Date_____

Tell about the lie(s) you have lived because you felt you had no other choice that no one knows about.

Because I Felt I had No Other Choice

Today's Date_____

My Most Embarrassing Moment(s)

Tell about the most embarrassing moment(s) you ever had.

In Life

Today's Date_____

In School

Tell about the most embarrassing moment(s) you had in school.

Today's Date_____

With Family

Tell about the most embarrassing moment(s) you have had with your family.

Today's Date_____

At Work

Tell about the most embarrassing moment(s) you have had at work.

Today's Date_____

In Public

Tell about the most embarrassing moment(s) you have had in public.

Today's Date_____

Tell about the most embarrassing moment(s) you have had that was out of your control.

That Was Out Of My Control

Today's Date_____

Tell about the most embarrassing moment(s) you have had that was within your control.

That Was Within My Control

Today's Date_____

Party

Tell about how you wanted to have a party but never could.

Never

Today's Date_____

All The Time

Tell about all the parties you had or went too.

Today's Date_____

With Family

Tell about the time you parties with family.

Today's Date_____

With Friends

Tell about how you partied with friend(s).

Today's Date_____

When I Was Jealous
Of Spouse(s)

Tell about the time you were jealous of your spouse(s).

Today's Date _____

Of Children

Tell about the time you were jealous of your children.

Today's Date_____

Of Family

Tell about the time you were jealous of your family.

Today's Date_____

Of Friend(s)

Tell about the time you were jealous of your friend(s).

Today's Date_____

Of Others

Tell about the time you were jealous of others.

Today's Date_____

When I Thought Others Were Jealous Of Me

Tell about the time you thought your spouse(s) was jealous of you.

My Spouse(s)

Today's Date_____

My Children

Tell about the time you thought your children were jealous of you.

Today's Date _____

My Family

Tell about the time you thought your family was jealous of you.

Today's Date _____

My Friend(s)

Tell about the time you thought your friend(s) were jealous of you.

Today's Date_____

Others

Tell about the time you thought people you didn't even know were jealous of you.

Today's Date_____

Vacation(s)

Tell about the best vacation you ever had.

Alone

Today's Date_____

With Family

Tell about the best vacation you ever had with family.

Today's Date _____

My Likes & Dislikes

Tell about the things you like or dislike about yourself.

My Life

Today's Date_____

My Spouse(s)

Tell about the things you like or dislike about your spouse(s).

Today's Date_____

My Children

Tell about the things you like or dislike about your children.

Today's Date_____

My Family

Tell about the things you like or dislike about your family.

Today's Date _____

My Friend(s)

Tell about the things you like or dislike about your friend(s).

Today's Date _____

My Job

Tell about the things you like or dislike about your job.

Today's Date_____

With Others

Tell about the things you like or dislike about others.

Today's Date _____

With The World

Tell about the things you like or dislike about the world.

Today's Date_____

Politics

Tell your view(s) on the Democratic Party.

Democrat

Today's Date_____

Republican

Tell your view(s) on the Republican Party.

Today's Date _____

Voting

Tell why you voted the way you did or if you did not vote at all, and why.

Today's Date_____

Hobbies

Tell what your hobby(s) are and about how you started.

Hobby(s)

Today's Date_____

Collections

Tell about your collection(s) and what you would like your family to do with them.

Today's Date_____

Collectables

Tell about your collectable(s) and what you would like your family to do with them.

Today's Date_____

That Thing I Always Wanted To Say

Go ahead and say that thing you always wanted to say.

To Family

Today's Date_____

To Friends

Go ahead and say that thing you always wanted to say to your friend(s).

Today's Date_____

To Spouse(s)

Go ahead and say the thing you always wanted to say to your spouse(s).

Today's Date_____

To The Law

Go ahead and say that thing you always wanted to say about the law.

Today's Date_____

Today's Date_____

To The World

Go ahead and say that thing you always wanted to say to the world.

Today's Date_____

That Thing I Always Wanted To do

Tell about that thing you always wanted to do with yourself.

With Self

Today's Date_____

With Family

Tell about that thing you always wanted to do with your family.

Today's Date_____

With Friend(s)

Tell about that thing you always wanted to do with your friend(s).

Today's Date_____

With Spouse(s)

Tell about that thing you always wanted to do with your spouse(s).

Today's Date_____

Taxes
Tell how you feel about taxes.

How I Feel

Today's Date_____

How I Believe

Tell what you believe about taxes.

Today's Date_____

Tell what you think should be done about taxes.
What I Think Should Be Done
Today's Date_____

Racism

Tell what you feel is good about racism.

The Good

Today's Date_____

The Bad

Tell what you feel is bad about racism.

Today's Date_____

And The Ugly

Tell what you feel is ugly about racism.

Today's Date_____

Computers

Tell how you feel about the computer in your life.

Computers In My Life

Today's Date_____

Tell how you feel about the computers impact on the world.
Computers In The World
Today's Date_____

How I Feel About Myself As A Man

Tell how you feel about yourself as a man.

Today's Date_____

Woman

Tell how you feel about yourself as a woman.

Today's Date _____

Military Service

Tell about your time in the military.

Service
- Army[ ]
- Marine Corps[ ]
- Navy[ ]
- Air Force[ ]
- Coast Guard[ ]
- Reserve/Guard[ ]

Rank/Medal(s)

How I Feel About Myself As A

A Sister

Tell how you feel about yourself as a sister.

Today's Date_____

A Brother

Tell how you feel about yourself as a brother.
Today's Date_____

A Cousin

Tell how you feel about yourself as a cousin.
Today's Date_____

An Aunt
Tell how you feel about yourself as an aunt.
Today's Date_____

An Uncle

Tell how you feel about yourself as an uncle.

Today's Date_____

A Niece

Tell how you feel about yourself as a niece.
Today's Date_____

A Nephew

Tell how you feel about yourself as a nephew.
Today's Date_____

A Grandmother

Tell how you feel about yourself as a grandmother.
Today's Date_____

A Grandfather

Tell how you feel about yourself as a grandfather.
Today's Date_____

A Great-grandmother

Tell how you feel about yourself as a great-grandfather.
Today's Date_____

A Great-grandfather
Tell how you feel about yourself as a great-grandfather.
Today's Date_____

A Friend

Tell how you feel about yourself as a friend.

Today's Date _____

A Co-worker

Tell how you feel about yourself as a co-worker.
Today's Date_____

Homosexuality
What I understand about homosexuality.

What I Understand

Today's Date_____

What I believe about homosexuality, from my understanding.

What I Believe

Today's Date_____

Sports

Tell about your achievements in sports.

My Achievements

Today's Date_____

Professional Sports

Tell about your view on professional sports.

Today's Date_____

My View On Global Warming
What I believe about global warming.

I Believe

Today's Date_____

Disbelieve

Tell why you disagree with global warming.

Today's Date_____

My View On Crime

Tell what you think about crime in your town.

What I Think

Today's Date_____

What I Know

Tell what you know about the crime in your town.

Today's Date_____

What I believe

Tell what you believe about crime in your town.

Today's Date_____

Book(s)
Tell about the book(s) you have read.

Read

Today's Date_____

Recommend

List the book(s) you would like to recommend others read.

Today's Date_____

Vehicle(s)

Tell about the first vehicle you ever owned.

First

Today's Date_____

Wanted

Tell about the vehicle you always wanted.

Today's Date_____

Loved

Tell about that vehicle you loved.

Today's Date_____

Television

Tell about your favorite television show, and why it was your favorite.

Favorite

Today's Date _____

Worst

Tell about the television show you thought was the worst.

Today's Date_____

Loved

Tell about the television show you loved and why.

Today's Date_____

Actor(s) - Actress(s)

Tell about your favorite actor(s) - actress(s).

Favorite

Today's Date_____

Worst

Tell about the actor(s) - actress(es) you though were the worst.

Today's Date_____

Loved

Tell about the actor(s) - actress(s) you loved.

Today's Date_____

Outer Space
What I believe about space travel.

Space Travel

Today's Date_____

Aliens

What I believe about aliens.

Today's Date_____

What I Believe

What I believe about outer space.

Today's Date_____

Music

Tell about the music you love.

Love

Today's Date_____

Hate

Tell about the music you hate.

Today's Date_____

Movie(s)
Tell about your favorite movie(s).

Favorite

Today's Date_____

Worst

Tell about the movie(s) you think was the worst.

Today's Date_____

Loved

Tell about the movie(s) you loved and why.

Today's Date_____

Hated

Tell about the movie(s) you hated the most.

Today's Date_____

Shopping

Tell why and what you love about shopping.

Love

Today's Date_____

Hate

Tell why and what you hate about shopping.

Today's Date_____

Addiction

Tell about your shopping addiction.

Today's Date _____

Favorite Designer(s)

Tell about your favorite designer dress or suite.

My Favorite Dress - Suite

Today's Date_____

Tell about your favorite designer shoe or handbag.
My Favorite Shoe or Handbag
Today's Date_____

Favorite Weather

Tell how you feel about winter.

Winter

Today's Date_____

Spring

Tell how you feel about spring.

Today's Date_____

Summer

Tell how you feel about summer.

Today's Date_____

Fall

Tell how you feel about fall.

Today's Date_____

My Favorite Food

Tell about your favorite food.

Favorite

Today's Date_____

Worst

Tell about the worst food you ever had.

Today's Date _____

Love

Tell about the food you love the most.

Today's Date _____

Hate

Tell about the food you hate.

Today's Date_____

My Pet(s)

Tell about your favorite dog(s).

Dog(s)

Today's Date_____

Cat(s)

Tell about your favorite cat(s).

Today's Date_____

Other

Tell about your favorite pet(s).

Today's Date_____

My View On Life Support

Tell what you believe about life support for yourself.

For Myself

Today's Date_____

For Others

Tell what you believe about life support for others.

Today's Date_____

Death Penalty

Tell what you believe about the death penalty.

Death Penalty

Today's Date_____

My View On Death

Tell your view on death for yourself.

For Myself

Today's Date_____

For Others

Tell your view on death for others.

Today's Date_____

What I believe

Tell what you believe about death in general.

Today's Date_____

In My Own Words

Use this section to add info that is not mentioned in the chapters provided, that you want to say in your life book.
Today's Date_____

What I Think Will Happen After I Am Gone

What I think will happen to my family after I am gone.

With Family

Today's Date_____

With The World

What I think will happen to the world after I am gone.

Today's Date_____

My Last Will And Testament

Last Will and Testament
Of

_____

Today's Date _____

I, _____, Presently residing at City of _____, in _____ County, State _____, being of sound mind, declare this to be my Last Will and Testament. I leave _____ _____ as the executor of my estate.

Notary

```
┌─────────────────┐
│                 │
│                 │
│                 │
│                 │
│                 │
│                 │
│   Notary Seal   │
└─────────────────┘
```

Date_____/_____/20_____

Signature_____ Notary Signature_____

Witness_____

This is my last will and testament, in my own words.

Signature_____ Date____/____/20___

Place photo here

Last Entry
Day_____Month_____Year_____

Live life, Do life, Shear Life, Give life

Save life, Endure life, Welcome life

Start life, Feel life, Breath life, See life

Teach life, Touch life, Honor life

His life, Her life, Their life, Our life

My life, Some life, Is life, Through life

In life, Be life, Good life, Great life

About life, Enjoy life, Pursue life, A life

Think life, Do life, Beautiful Life

Interesting life, Guild life, Amazing life

But most of all **<u>Love Life</u>**!

The End

About The Author

**Vickie Smith lives with her husband Larry Smith, Head Pastor of the Lighthouse of praise Kingdom Center and their daughter Jessica in Indianapolis In,
She can be reached by E-mail at
PastorsSmith@sbcglobal.net**

www.ingramcontent.com/pod-product-compliance
Lightning Source LLC
Chambersburg PA
CBHW081412230426
43668CB00016B/2217